Quantum Shift

Alien Contact

Sylvester Ashcroft

Prologue – Event Horizon

In the beginning there was man, or was it a boy...and his pikachu! Oh scrap that wrong story...let's go back to the beginning, where it all started for me, when I was but a young boy...I always felt my father's watchful eye on my back, as he watched me work on my little projects. He always questioned me on the relevance to my studies, feigning ignorance I knew he was smarter than that.

He worked as an aerospace engineer for most of his life, rocket science was his day job or so he told me, that's what got me into the job I eventually found myself in, as part of my lifelong passion for exploration...through time and space. It wasn't until I was 12 years old, that I really felt awoken to my thoughts, the elusive man who haunted my dreams, a shadowy figure watching over me, who or what was he exactly?

I thought it was a ghost at first, until I saw a psychologist at the age of 15, as the visions grew worse, he said that somehow I had temporal amnesia, basically meaning 3 years of my life were missing from my memory, and my brain was trying to fill in the blanks.

The shadowy figure was most likely my father he told me, guiding me back to where I once was...my mother died during childbirth, so I was an only child, but my dad never gave up on me, not for one second, I couldn't ask for a better father if I tried.

The truth is my father was a saint, up until he died just a few years after his retirement, he told me; 'make your dreams come true, do me proud, you always have, I couldn't ask for a better son.' He then passed away right before my eyes, as I held his hand yelling no! NOOOO! They turned off the life support, and I knew from that moment onwards, I had to find a way to fix the wrongs I had done...

I spent all my time studying everything around quantum physics, and started drawing up my thesis for interspacial time travel. I submitted it to CERN, but my findings were rejected, the only person who would listen to me was my college professor, who was also my mentor of 16 years.

He was also my godfather...my father's lifelong friend, the closest thing I had to family. He taught me about Math when I was only 6 and he saw my potential, and saw how hard my father was with me, so he took me to the park, and did things my father would never let me do, he let me be a kid again.

I know why my father was like that, he only wanted what was best for me, but what was best was not necessarily what I wanted.

I was only 24 years of age when he passed away, I had no other relatives to speak, apart from my auntie who moved to Spain about 12 years ago. I became a recluse after that, completely absorbed with my research into chrono dynamics, only going out fresh air, every few days to clear my head. I was lucky the university was willing to fund my research, and after 3 months the project was up and running, that's where my story begins here...

Chapter 1 - An Object Of Time

You cannot defeat me, I AM IMMORTAL! I awoke with a start, I was dripping with sweat, what the hell happened, was I having another one of those lucid dreams. That shadowy figure from my childhood had returned to haunt me, with it's wicked red smile, and it's...it's one eye staring down at me. It was about 8 foot tall, and had arms as long as tree branches, with legs that seemed to pierce the very earth which it stood, as it stared down at me, I could feel it's presence closer than ever...it was coming for me, but I was not afraid!

I had to continue work on the machine, to stop this fiend from killing me, from the inside...I needed to finish my research before it drove me insane...I drove to work on the normal morning commute.

I whacked the coffee machine a couple of times asusual, took my first gulp of the day, to make myself feel somewhat alive, and went back to work. I said hi to the lab techs, who had been working tirelessly overnight, to get us up and running, as the last test caused a massive blackout across half the city, but that was normal...as we're tapping into multiple unknown regions of time and space.

It was bound to require a massive amount of energy, which is why we expanded our research into nuclear fusion, they had already made some progress towards a sustainable fuel cell, but they told me it was 20 years away from being manufactured.

I told them we needed it done within a month, so we needed more funding to be able to do that, otherwise I would have a seizure before then! I was the head engineer on the project, we were always the biggest teller on accounting, and got multiple raised eyebrows from people who little about our work. The kind of research we were doing could change the world, but they were only interested in their soap operas, and Instagram.

I didn't carry around a mobile phone, as I felt like they were mind numbing caskets of radiation, even in small doses it still effects brain cells, so I kept well away from that. I don't drink nor smoke, but if someone offered me a decent steak, I'd happily drive them home afterwards.

The sensible option would be to retire with a nice paycheck once our research was complete, but I didn't plan on handing over my life's work so easily, they would have to challenge me first...and win...which would never happen, as I was always one step ahead of them at every turn, as they were too preoccupied by their own desires.

As I hammered out the specifications for the machine I felt the everlooming presence of the director, the elusive administrator who had funded this whole project, in a similar way to M from James Bond. I did not know his name I daren't ask, he was as elusive as he was blunt, stating simply progress, then normally good work...continue.

I simply considered him to be a sack of shit for the most part, but he did save me buying new clothes for at least a fortnight after I started, as almost everything I owned was covered in black turd.

As the engine fuel we were testing to start with, had a tendency to explode all over me, like a cow that suddenly decided this was a nice spot, to drop a mega crapper, as everytime we drove past a field, it reminded me of the smell my father left in the bathroom, which waned all the way into the front room, after gobbling down his cereal, lovely stuff!

It is always nice to bring back memories of your childhood, preferably not including the one's where you sneak into the kitchen, at 6 o'clock in the morning because you can't sleep, wake up the cat, and get the slipper, as he won't stop meowing.

The blackouts had become more persistent as of late, meaning I needed more regular breaks. I would find sometimes wake up, and find myself sitting in a corner shivering, damp with sweat, like I had been sitting there for hours, rigid with fever.

The doctors couldn't find anything wrong with me, even my cat Copernicus couldn't understand what was wrong. He'd just sit there looking at me, like I had something on my face, he couldn't understand me at all, I wish cat's could talk, as I stroked across his head, wondering what to do about this. I could barely go outside, I kept myself locked in my office.

My good friend George was always there to help with my homework, but I often got sidetracked, I was more interested in playing with aeroplanes than my homework, I then moved onto making pressurised rockets in my backgarden, and trying to turn the water hose into some kind of high pressure blaster, which I fired balls at the neighbours fence for fun.

It wasn't until I was about 25 that my thinly veiled sheen of moderately greasy hair, and out of date fashion sense eventually got me noticed by one professor of mine, whom I approached about a project I was theorising which involved quantum space manipulation, in order to teleport matter from one point to another, by changing it's atomic frequency, in a similar way to tug of war, whereby alleviating the pressure of one object from another would cause the velocity of the object to massively increase, causing it to propel forward at massive velocity.

This in concept proposed that matter was constantly in flux, and every atom within our bodies had a specific frequency at which they vibrated at, and if they could be tuned to a certain frequency, they could become static long enough, to transpose themselves from one point in time, and space to another.

This is where the great and wondrous adventure, or as I like to call it...my personal hell, great balls of fire! This time hopefully it won't explode in my face, again...

3 years later...

I flick through the TV channels to try and find something interesting, and come across the news, and decide to leave it on a moment, to see what's going on outside our ivory bunker.

Welcome to DNN(Daily News Network), the 12 o'clock news will start soon, but first a few messages from our sponsors! (cuts to static)...test signal bzzzzzzzzzzt...(audio only subtext)...

Citizens of earth...do not be afraid, for now is the time of your redemption...you have lived in the darkness for too long...but now WE have returned...

If you wish to seek forgiveness it is too late, the time of your redemption is almost at hand! (a low rumbling begins across all the major cities) you may think that is an earthquake you are experiencing, look outside your windows...you will see the truth, you are not alone, you were never alone...

I pulled back the curtains and saw reflective discs levitating down from the sky above...you have ran out of time, the age of humanity is over, the age of the silver blood has begun, you cannot resist...or else you will be terminated.

The military sends in jets to attempt to take down the alien spacecraft, but their missiles are being interfered with, and are forced to retreat as their weaponry is reprogrammed and used against them. We have studied your weapons...they are inferior to our technology, which is 1000's of years ahead of anything you could understand, you human's are fragile creatures, we are not affected by such weakness to space.

If you retreat now maybe you too could ascend to become like us...the creators of your race...What the heck...I said as I dropped my coffee, a lab tech rushed over and asked me 'Are you okay?' I said 'look at the TV...what in god's name, ah my ears!' I couldn't hear a thing, but she was bleeding out of her eyes, ears, and nose, I turned off the television, and it seemed to help somewhat, I had to call a medic in to take her to the hospital, could it be they have been communicating with all this time, and why am I not affected by the rogue signal, they are distributing across the airwaves?

Chapter 2 - 10,000 BC

In the beginning there was an entity which consisted solely of grey matter, a literal hive mind of globulous fluid, they eventually developed sentience, and became the prime species in the universe, travelling the stars to find the perfect extension for their race, that is when they discovered earth, long before humans ever existed...They believed by the year 2050 AD most of western civilization would be reduced to ruin due to war or famine, but they were wrong...thousands of years before humans began to thrive, as an intelligent, and well educated race of sapient creatures, we long held the belief that there was life elsewhere in the universe.

The Mayans took this very seriously even basing the locations of their cities on constellations in the night sky, even going so far as to ritually sacrifice their own people, in the name of their gods. One of the most famous among these was Quetzalcoatl, said to be a man with the head of a serpent, but could it have been the case that this god wasn't just a myth, and maybe even the Egyptians themselves had encounters with lifeforms from outside our solar system, which is how they constructed the pyramids thousands of miles apart, but with similarly shared history, shrouded in mystery.

It soon became apparent however that they had been observing us for a very long time, and were simply waiting for the point, where our race had reached their highest potential, so they would be ripe for harvesting as seeds of a new race, of smarter stronger creatures.

In the events that led up to this day, I kept a diary a log if you will of the events as they occurred, I write this postscript of this day in which the invasion began, and so in my own words I describe to you, the fall of the human race, to beings much smarter, and older than anything we could imagine, with technology thousands of years ahead of our own, this could also be the case that the catastrophe which destroyed the dinosaurs over 35,000 years ago could have been engineered, to cultivate the seeds for humanity to be born.

It was merely a process of natural genetic engineering, through a process of forced evolution, similar to how sea life can survive submerged absolute darkness, under immense pressure, which no human could survive for more than a few seconds, without their ribcage being crushed like being hit by a skyscraper head on.

It is only now that I realised what I must do, I must put my research to the test, little did the government know I had my own prototype of the machine they had me working on, and I purposely sabotaged theirs by removing key components I needed for my machine, which were incredibly expensive and difficult to obtain, as they were specifically designed for space travel, and for high velocity matter transfer, such as the research in CERN.

The technology was based on experimental research into alien life, the united states had kept above top secret, after a team of excavators uncovered the Sumerian ruins, they were camouflaged from google earth, and all knowledge of the subject was deleted from memory, otherwise any information spoken or otherwise would be classed as treason against the us government. In 1976 the US Military began research into experimental, long range submarines, which would use undersea pipe networks, to travel vast distances, across the ocean in a matter of seconds, using a wind tunnel concept, combined with the Sumerian technology, to create an interspacial tunnel, which would warp the space around it, causing it to be coated in a thin layer of sulphuric oxide.

This proved highly unstable however, claiming many lives, especially of those who were directly involved in manufacturing the power conduits, and the deep sea tunnel systems. They were infected with something akin to the black plague, they called it quantum flu, skin samples examined under the microscope of the areas effected, were thousands of years older than the surrounding tissue, which was simply impossible, but was due to the massive amounts of quantum radiation created when the two polarising caps were released, sending the submersible from one end of the tunnel to the other, radiating for thousands of miles, harming anyone nearby including the submersible crews, who were remotely tracking the trajectory, and speed of the craft deep under the ocean.

The tunnels themselves were lowered via cranes along the ocean floor via incredibly strong cables, capable of holding bridges above water. The cables themselves were concealed in highly secretive locations near deep embankments of water, with cables fed through local power stations, eventually leading to the major power grids nearby. This approach led to mass blackouts across the west coast, with panic and hysteria ensuing most of the western united states.

It wasn't until the distant future that research into this phenomena was reopened, with the introduction, of quantum field technology, which allowed for practical applications of this ancient technology, in the modern day, using modern day technology, without the need for primitive power sources. This caused ripples through time however causing irreparable damage, the region which these tests were conducted in an area east of the USA, known to us as the Bermuda Triangle.

80 years later...

The design wasn't perfect but it will have to do, as I only have 3 days before the invasion comes full scale. As the scout ships have already started descending, but at the moment the atomic clock, can only transfer my atomic data, approximately 73 hours 5 minutes and 23 seconds into the past, or possibly...into the future, but I daren't try that yet. I'm simply using it as a means to extend the time, I can work on the machine, before the invasions begins full scale, and I must continue, otherwise the human race as we know it, may cease to exist. We could be replaced by those soulless lifeforms, which seek to replicate our race as their own...
'In the beginning there was man, in the end there was only them...' Arthur T. Stanton PHD

I wake up to see flickering spots of light scatter across my vision, the back of my eyes burn with a fierceness like that which I have never seen before, it is then that in a fraction of a second, I open my eyes completely and see the world in flames...strange alien beings in machines comb across the woodland, until one focus' on me directly, but how could this be. I was only eight years old when I had this vision for the the first time, and over the years I came to the realisation something must be coming, and I was just somehow more aware of it than other people around me, and we are most definitely not alone in this universe, it's just a matter of time...

Chapter 3 - Creatures Of Light

I awoke once again to the morning sun, and pondered on the wonders of our sight, our eyes suffuse light in such a way, that we can interpret shapes and colours, in a way unique to most creatures on earth...when combined with our cognitive abilities, we are incomparable in the complexity of what we see, and understand about our universe, compared to most if not all creatures that exist in our universe, apart from the human race.

The theory however is that considering the age of our universe, realistically speaking there must be some form of intelligent life, other than human beings in this universe, whether they are on par, or more advanced than ourselves, that begs the question, are we really alone?

This is perhaps not as infeasible, as we might first imagine...scientists disagree on the exact qualities of what an alien lifeforms would consist of, but most agree on the basis qualities of life existing in our universe, e.g. a source of heat, natural light, and water are essential, and possibly the ability to absorb oxygen, from natural gases on the planet's surface, similar to carbon based lifeforms, such as in the same way that trees absorb carbon dioxide and release oxygen as a means to sustain life as we know it.

This is only a very fragile sense of an ecosystem which can support life, but is the basis for most of our research into extra-terrestrial entities which might exist within our universe, which we are a very small part of.

I do some very sensitive work with the military, and as of 3 years ago, we have begun working on a quantum entanglement device, known as the HDFG - Hyper Dimensional Field Generator, which was designed on the basis that it would be possible to bend light in such a way, that matter could be physically shifted through space and time, by reducing their mass to just above 0 then physically shifting their mass through space and time, then reconfiguring it, along with the machine at a different point in space and time.

This was deemed impossible for the longest time, but with breakthroughs in technology such as the work at CERN, it was believed it could soon be a reality. We designed a machine as part of this process, a quantum computer designed to specifically render an environment in 3D space where we could test, potential methods of quantum shifting, without actually causing physical harm to ourselves in the process.

We tried 1000's of different variables, and only one of them was successful, but it required a live subject, so we weren't able to test, as the machine required us to build the rest of the machine.

This fundamentally could be classed as a time machine of sorts, as the machine could accurately predict the future, albeit not that of future world events, but fixed variables within the confines of our lab, but it saved 100's if not 1000's of hours of man power in research, through continuous virtual iteration of our machine.

Chapter 4 - The Core

The device was powered by deuterium a naturally occurring gas in the earth's ocean, which was syphoned using a special deuteron syphon, as part of the planning process for creation of the machine, as funded by the military, so that it would have a significantly longer lasting power source, than most nuclear fusion reactors, which were designed today to be much larger, with significantly higher fallout. Through hydrogen degradation, they managed to create a sustainable means of crafting deuterium by pouring water into a cooling tank, situated on the back of the machine, which would then be syphoned off using a hard light laser, to exert the oxygen atoms from the hydrogen atoms within the water. Thus creating a gas pocket, which would be fired from the exhaust of the machine, creating oxygen within a high pressure vacuum similar to that of a submarine.

In reality though physically travelling through time is a much more complicated endeavour, since humans have a tendency to die, when exposed to high levels of stress on our bodies, being known to seize when we receive high doses of electricity, or if our minds our disconnected from our bodies for too long, the stress can also lead to heart failure, as well as other electrical impulses which are needed as human beings to keep us alive. We hoped that at the end of this, we would find a way to successfully quantum shift safely, without ending up on the other side as a pile of hot goop, like a microwave meal that unfortunately exploded as you left it in too long.

If we could find a way to shift our atomic mass through time and space in such a way, that we reform exactly the same way on the other side, without killing ourselves in the process, or smelling like a bucket of KFC, cooked through in the same way we were reformed, or with our arm poking out of our skull, and one leg attached upside down to the side of our bodies, this is the unfortunate side effect, of some failed experiments into time travel in the past.

The problem was as I drew ever closer to the machine we were creating, I soon came to realise I couldn't allow them to finish the machine, as this was too powerful, for any one nation of people to control, so I purposely tampered the results of the tests, creating false positives, radically different from the actual end result, since I was the lead engineer I was the only one who had direct contact with the machine, as I was the only who was safety briefed, and security checked 3 years prior to getting this lucrative position, with a decent pension at the end of it.

I couldn't allow success to be an option, and eventually just as I hoped the project was shut down, but little did they know, I had copied the schematics of the machine, and I had been working on a similar device for month's now, albeit one that works, this was fortunate however as if they realised what they had created, the world could be a very different place indeed...

'we are at great risk of an existential event, we can only hope it doesn't occur...' Arthur T. Stanton PHD

Chapter 5 - The Syphon

30 days later

I worked around the clock since the day of the incident, I realised that I could not wait for them to come to me, so I had to do something. The chronometer is a device which inhibits the time flow around the device, using an enhanced version of a standard wall clock, using black quartz I bought, thank god for eBay I told myself. This was to make certain that the machine that had a reading an order of magnitude more accurate. The machine was close to completion, I stroked the dust off the atomic clock, in it's catatonic state of quantum resonance.

It was the heart of the machine, a fuel source was really hard to come by however, and it took me several days to theorise a long term solution for this. I realised as I was pondering this, that I had accumulated black substance on the soles of my feet, from repeated shifting through the timeline, this seemed to be a side effect of repeated travel through time, as your mind accumulates decades, but your body doesn't age more than a few days at a time. I needed to study the side effects more closely, as for now it seems to be just a tarry substance accumulating on my feet, as far as I can tell there are no other side effects which I can see.

I constructed this device in the hope that it would prevent this invasion from ever occurring, but I feel it won't be enough, this race is much more advanced than anything I could have imagined, one jump I wandered around too long, and was almost seared as a scout vessel was approaching on the horizon, so I quickly had to run inside, and had to activate the machine, before I became their next meal, or so I hoped.

I do not know what they intend to do with the survivors, but it definitely can't be good, they surely can't have accounted for someone, circumventing their own technology against them, by rolling back the clock so to speak before the invasion ever occurred. I know one thing for sure, if anything I must fight, for everything that is right on this earth, otherwise there will be nothing left to fight for, and the memory of humanity, will be reduced to nothing but dust, like grains of sand blowing in the wind, as the new rulers of earth consume us all...

Chapter 6 - Entrapment

83 years ago I designed a machine, that could slow down time to a crawl, by destabilising the atomic structure of the space around it, and slowly down the particles, to the smallest possible percentile.

I then accelerated this process and found a way to slingshot myself backwards, and forwards through time, using a process I called Quantum Shifting, it was like transferring data across the internet. My quantum signature would be displaced a fraction of a second after I activated the machine, the problem is I soon realised, that time itself had become damaged in the process, and as I write this diary I am trapped here in the year 2136.

I cannot find a way home, as time itself is tearing itself apart, so I am going to use my machine one last time, the final version of which I had created, a time watch designed to attach to the wrist of the wearer, which would map their quantum signature, using a system of sonic vibrations, which would destabilise the atoms to the subject on the base level, allowing them to desynchronise from time, the problem is...It was a one way trip, and it only had enough power for less than 10kg of atomic mass, so I'm sending this capsule to you, myself in the past, warning you to never create this machine, or....to complete my work, and stop global Armageddon, which I bore witness to almost 60 years ago, which was why I created this machine.

I never met the other versions of myself as they were from my past, before I created this version of my machine, and each time I did so, time splintered, and the previous version of time ceased to exist, which eventually led to the wholesale mass destruction of the human race as we know it.

I do not permit this, but if necessary I enclosed a stone key, you will know where to go when you need it. This is the fail-safe for my past self, a reset switch so to speak...the critical axis, at the focal point of time and space, hidden under the rocks of Stonehenge, look for the pillar with a small rectangular depression on it, this will open the way.

Inside you will find an access point, to a place outside of time, at the edge of space. You will find the secrets for mankind's downfall, a documentation of everything that has happened up until now, unaffected by the destruction of the time streams, powered by a network of solar energy relays, atop the pillars which I set up long before Stonehenge, was even discovered by modern day humans.

This is why you can never tell anyone about your mission, otherwise time itself may be at risk of collapsing as we know it, only the ethereal barrier that protects that place will survive, but everything around it will be destroyed, and you will be trapped in the past, so don't forget your mission.

I started the camera and spoken clearly and calmlyinto the lens, my name is Jacob Carter born October 16th 2034 and this is my story, a story about the end of the human race, and the rise of something....much deadlier than anything we could have imagined, we are most definitely not alone...

Chapter 7 – One last trip

October 19th 1837

The pills weren't helping me, the sickness came in waves, I kept coughing blood until I felt like I was going to pass out, I knew this wasn't good, the quantum shifting was taking it's toll on my body, I couldn't keep doing this, but the machine was almost complete, each time I shifted I found another component I could use, antiquity seemed to have a lot of unused or otherwise lost technology, whatever happened to antikythera for example.

I would have to make another jump to find out, but that might be one of the last parts I needed, but I'd first need a particle scanner for that from the future, as they won't be invented for another 250 years at least. I went down to the bar, shaking I climbed on the barstool and rested my head on the wood, like it was a pillow.

Hey buddy you don't look so well, let me buy you adrink, a whisky on the rocks for this good sir! I looked up and saw a friendly smiling gentleman sitting some 3 feet away, with a grin as wide as the gap between my thighs. I had to remember my mission, so I smiled politely, and said thanks! what's your name buddy? I noticed the military uniform he was wearing, I was wondering if he was leaving or if he had just come back.

He got up and walked towards me, throwing me off slightly, corporal lance armstrong, nice to meet you sir! He outstretched his hand towards me, his impeccable fingernails trimmed to within the edge of his skin, probably to keep the blood out of his nails.

He was making me feel uneasy, so I extended my hand in response Jonathan Barns I replied, the best I could come up with at such short notice, I remember seeing a letter in the lobby addressed to a Jonathan Barns, I hope he isn't here otherwise I'd be rumbled. nice to meet you sir, he says politely, now what brings you here, business or pleasure might I ask? a bit of both I reply...

I then grit my teeth and tell him a lie saying the wife is having a grand old time at the ministry ball, but I came down with a bad case of fever, so I couldn't accompany her there, as it wouldn't be polite to cough on all the dignitaries there. I then said where are you posted then? He then said ah! this thing, I just enlisted, got my uniform today, haven't you seen the signs, he points at a nearby poster.

I mock looking interested towards it raising my eyebrows slightly, ah I see. He then replies yeah uncle Sam, needs men and women for the front lines, what kind of job are you into sir, you look like a doctor if I may say so. I then nod in approval, feigning a sense of knowledge in my craft, yeah I swung by as there was a fever going round, and I seem to have caught it myself, whilst doing the rounds in town, pretty nasty stuff, people coughing up blood and all sorts, it seems to be worse in the elderly.

That's sad to hear sir, well hopefully you get better soon, you better get some rest after you finish that drink, nice meeting you sir, he nods approvingly as he leaves some money at the bar, the barkeep wasn't looking as he took the glasses and left the money on the side, so I took the money and made my way out of the building, I could hear someone yelling behind me; hey get back here! but I was gone before they could catch me...

Chapter 8 - Going home

October 16th 2037 present day

I really needed to take a leak, so before making my notes, I went immediately to the restroom, before typing up my findings. I looked at my face, and I noticed the plague was getting worse, the medication was slowing it down, but my cells were slowly dying from what I like to call quantum necrosis, or time syndrome. It seems to happen from repeated trips through the time stream, dam it! I say as I stub my toe on a rusty nail protruding from the floor. I spray it with some disinfectant, and wrap a bandage around the open wound.

The gash is bleeding quite severely, as I feel my toe swelling to the size of a small grapefruit, but I can't stop I'm so close now, just a few more components and the machine is done.

The time capacitor is almost complete, a makeshift bomb designed to disintegrate all organic life completely from time itself.
It's a hunk of junk now, but in theory this should slow down time long enough, to completely disintegrate them from inside their own mothership. A convenient if not somewhat contrived plan, which would require a lot of pieces to be in the right place, at the right time in order to succeed.

In order to create their machines, they fused scrap metal, using a quantum accelerator powered by a nuclear reactor, and created interspacial anomalies to allow themselves to travel across vast distances in space, but like miniature black holes. They were around 100's of thousands of years before humans were even a grain of sand on the wind. They utilised cloning in order to transfer memories from one generation to another, to massively accelerate the growth of their race, building an empire within a 1,000 years or so of accelerated growth, and hormone culturing.

I believe the destruction of the dinosaurs was an experiment gone wrong, and the only reason humans exist at all, is because of the technology they used to destroy the dinosaurs accelerated the development of new life on earth, as they came from 1000's of light years away, in a system with no sun, for the first 100 years most of their race was blind.

It wasn't until a new star began to develop just outside their planetary system, guided to the light, their growth accelerated drastically, their long cylindrical bodies, then develop limbs similar to that of early humanoids but they were celluloids, creatures made from a gel like substance, which could develop without light or water.

They eventually found nourishment, and because of this, their lifespan expanded way beyond that of the humans, as their bodies hardened to form a hard diamond like surface, to become denser in the harshness of the morning star. They developed ways of creating food, and harnessing basic solar energy, although muted by human standards it was a start towards them developing a sustainable source of energy, in order to grow as a race.

It wasn't until the early 1800's that they were able to blend amongst the humans, before that they were studying them as a race, and created a complex series of symbols, as a way of communicating with other members of their race, which were hidden amongst the humans.

They could absorb human DNA, and replicate it within seconds, sometimes appearing as a grey goo like substance outside of water fossets, absorbing them whole. They did not deem us a threat until they saw the potential of human weaponry in the early 1900's so they began their attack abut 250 years in the making, and so it begins, to tell you the story of the fall of our race, we have to go back to when it began from the very end...

Chapter 9 - Hellscape

I accidently slingshot myself to Mt Vesuvius just after the eruption at Pompeii, the ground is still searing with molten lava, but luckily I'm outside the precipice of the molten inferno that completely immersed the city. I can see something strange gleaming through the fog of ash and smog, an alien ship is scanning the wreckage of the city, searching for something, why were they here? This could be why my atomic clock spun out of control, sending me over 200 years into the past.

I sneak around the water just outside the city's boundaries as I attempt to observe what it's looking for exactly. I pull out a handy pair of foldable binoculars I picked up on my travels and zoom in on the ship. It has some kind of mining laser trained on the peak of Mt Vesuvius just near the eruption area, maybe it wasn't a natural occurrence that caused Pompeii to be destroyed by this eruption, and it might be looking for something that crashed here, causing the eruption possibly.

It probes at a small hole retracting it's claws, as it pulls out a small glowing piece of rock, about 2 inches across and 3 inches across, which seems to be reflecting the light like a mirror, it appears to be some sort of diamond shaped prism, was it made from obsidian, or some kind of alien element, I wondered to myself, little did I realise a small drone was scanning the surrounding area, and found my standing there, staring eagerly at the ship.

I turn around after getting the feeling I'm being watched and jump into the water, the ship comes barrelling down the side of the mountain, rolling down like a boulder, something which I have never seen before, maybe this is a newer model sent from the future, or just a local variation for this mountainous area.

It surprisingly then submerges itself in the water, and a large retina appears on the front of the ship, with three black rings, and a large black pupil staring directly at me, it then retracts one of it's claws and attempts to grab me, but it doesn't realise I have created a weapon to combat their technology now.

A weapon designed from transient alloys, which are used to create their spacecrafts, as well as their own weapons, which I carefully transposed using high voltage electrolysis to separate the elements, and then forge them into a sword, using a high temperature 3D printer, which basically cooked the ingredients in a complex mould, using lasers to shape and define the features of the weapon.

It took several hours, and after that it had to be left to cool for around three days, then it had to be sharpened and polished, after 12 days the weapon was complete, and could slice through even the toughest of enemies hides.

It then grows more hostile after I slashed off one of it's claws, and draws out it's mining laser to attack me with, I notice wires attached to the inside of the capsule, and I have an idea.

It then begins to charge the laser after aiming it towards me, and I carefully dodge to the right, and swim rapidly around the back of the machine, grabbing hold of the lasers housing with my left hand, and slashing across the circuitry with my right hand, causing it to short circuit, sending the pilot into a panic, as it attempts to escape I slice open the hull, drowning him in the process, I then place small booster pads on the base of the ship, causing it to rise to the surface.

I put on a small gas mask I carry around with me, and investigate the ship. I find a small device within a protective housing which I assume was what it was looking for, it does appear to be some sort of prism, I also take a blood sample from the alien, using a diamond tipped syringe created especially for this purpose, and then I make my way back to the edge of the shore.

I activate the boost thrusters secondary functions, which causes a quantum anomaly to occur, destabilising the ship, and sending it into the void, along with a small portion of water, so that the existence of such technology doesn't destabilise the timeline.

I then reconfigure the portable computer on my wrist, as I set the coordinates to home, 3 days before the present day, so I can study what on earth, this small prism was going to be used for, and can find out what this alien's blood consists of, and whether I can replicate to improve my own self healing capabilities.

Chapter 10 – back to reality

His sole ocular lens focused down on the planet below, as a bullion of vessels gathered around the ship from the far reaches of space, cloaked with advanced reflective stealth technology, he was investigating the earth's defences, as his scouts disabled the united states nuclear warheads, by diffusing electromagnetic bombs inside the compounds where the bombs were held, having infiltrated them some 20 years+ before they were to be activated.

They were the perfect soldiers up to that point, as they were a literal clone of the officers which they copied, using the poly-organic grey matter, known as gris fatus or fatal goo. It wasn't until around 20 mins later that they received the signal, as he turned around and nodded towards the fleet commander, and an array of lasers were fired at the Whitehouse...the Pentagon...the United Nations...the House Of Commons...and the European Union itself...deemed as suitable targets for destruction.

There weaponry consisted of highly reflective mirrors, designed in such a way as to reflect highly concentrated beams of light onto their targets, turning them into a heat ray. They used the concentrated energy from a dwarf star to power their ships core, enhancing the laser energy further to the point of it being as precise as it was deadly.

There was nothing left as the blasts wiped away the remnants of the international defence forces, the CIA, FBI, NSA, and global defence councils had no defences for this systematic destruction of the earth's defences. The human race as we knew it was over, and the time of the alien's invasion had truly begun...

They only had one choice but to surrender, the ships initially took on the form of the standard UFO's we know of, then four pronged legs jutted from the sides of the ship, and the sphere rotated to form an eye, with a large laser which scanned for hostile life in the area, as it scoured the remnants of area 51 the scout ship incinerated numerous troops, as they came towards it, and tried to shoot at it's hull, it heated their flesh to over 200 degrees in a matter of seconds, disintegrating them entirely. The remaining soldiers, either attempted to flee, or were destroyed by the oncoming forces of the alien race.

I adjusted the quantum displacement field readying for another jump into the past, as the reading's on the telemetric scanner went off the charts, there was a surge in catatonic energy in the past, they must have accelerated their plans due to recent events with the reformation of the military in the future, but how could they know in the past?

Are they somehow linked through a hive mind, which can communicate across time and space, in a similar way to how telepaths can communicate across continents, possibly I suppose as this race seems to have advanced from a much more substantially harder environment than us humans, otherwise they wouldn't have been able to build ships which could float without thrust.

The field was up and running, as I set the coordinates, the space around the device rippled into action, attuning the quantum signature to the relative timeline, as my body shifted to the frequency of the new timeline, and I made the jump into the past but before the catatonic rippling just occurred, as I was getting closer to discovering how to defeat them, before they could enslave the human race as a whole.

Chapter 11 - Lost in time

September 21st 2053

I was working on a project for the military known as project time, they weren't exactly known for imaginative names, we were working on tactical time technology, in order to further the interests of future incursions, mostly against the Russian front-line forces, as their military had grown substantially since they overthrown the Kremlin some twelve years ago.

I was born to a family of farmers, but I was always interested in building things, my first invention was a box with mirrors attached, which was invisible from certain angles due to the way light reflected off it, camouflaging it from the environment around it.

I then moved on to making simple propellers, using tape an rc car motor, and some old engine parts for the fan blade, and the exhaust to give it some thrust.

It swung wildly from side to side, and had to be wired to a tractor for power, as I could not build a working battery, than was light enough to fly, but it worked although very basic, and it could only hover at about 3 feet off the ground, without flying wildly and hitting something, like my face.

I eventually entered the school science fair with an improved version of the flying device, but it was crude in design, and the other kids laughed at me, calling me nerd, and throwing food at my face, as I cleaned off my glasses, I told myself I would prove to them all, I was capable of much more than this.

I chucked the helicopter in the bin, but a teacher retrieved it, and kneeled down in front of me, and I will never forget what they said; 'don't give up just cause other kids pick on you, they are only jealous of your potential, never give up and you'll achieve great things.'

I never saw him again after that, as I watched intently as the young boy walked away, peeking through the fire exit doors, I realised that teacher was me from the future, so something more must be done here in the past, but what did I need to do? I retrieved a document off the floor, I didn't realise it was there before, it said in scribbled letters, meet me in the lunchroom at 12am come alone...

I realised it was 11:45am and the school fair wasn't over until 12:30pm so I wondered what this could mean, but I definitely felt curious enough to go check it out. I eventually find the lunchroom after getting lost, and almost stumbling past the headteacher's office, before I realised the door was slightly ajar, so I didn't want to be seen. A familiar voice said when I entered the lunchroom 'Why are you following me?'

As I turned around it was like looking into a mirror, as an older version of myself was staring back at me. He looked like a science teacher, an age away from where I am now. He stared at me for a moment, then responded, How can it be? I have been so far removed from my own timeline, that I now see myself standing right in front of me, how fascinating!

I feel somewhat confused, as the stranger circles around me, he says do not come closer! The same matter cannot occupy the same space otherwise neither would cease to exist!

He then slides a small watch towards me, this is the chronometer I have several of them, as I believe I may have caused myself a paradox by being here, you need to change the future, even if it means I do not exist in your timeline, you must do what needs to be done, but the chronometer in your timeline will only allow for one jump, until you find a suitable energy source, so only use it in an emergency, as you may need to It if they ambush you. Good luck to you my old friend!

He then phases from existence, as I make my way back towards the ship, cloaked from view inside a large closet, luckily nobody found it too strange to find a young man climbing out of a closet. As I head back home, questions shoot through my head, was that mysterious strange really me, or was it just a phantom was my future?

I couldn't know for sure right now, maybe we would cross paths again, and also this strange device he gave me, I wonder if I could figure out how it works, and maybe engineer a similar device of my own.

It will probably take months to complete a working prototype but if it allows me to quantum shift without the use of the machine, then it will save me days in repairs, as each jump takes it's toll on the machine, and it's only a matter of time, before it gives up completely, and I might be stuck in an alternate timeline forever, but I can't worry about that now, I must press on to the future.

Chapter 12 – Undead Whispers

I struggled to sleep as strange noises echoed through time, their voices were in my dreams, help me...PLEASE! They called out in the thousands, it must be because I keep crossing through the time stream that I am somehow connected with it, like it's literally a sentient creature of some description. I splash some water on my face from the filthy basin I call a sink, I then throw on my grey overcoat, and decide to go for a walk to the nearby town, as this quantum displacement is killing me inside, and I feel like I need a day to recuperate. I walk past a nearby pet shop, and I notice a small dog in the window seemingly staring at me from inside, I wonder why it's looking at me so endearingly so I walk inside.

I see the dog has a collar as it barks enthusiastically towards me, and wags it's tail excitedly, but no owner to be seen. I ask the lady at the desk about the dog, and they said a man left him here with a note, she passes me the note, and the handwriting is awful, but I recognise my writing anywhere, she said the man told her to wait for a man in a grey coat.

He would come in and ask about the dog, it belongs to him, even though he may not remember it yet. I looked down at the small creature standing before me, ad I soon came to realise why they call them man's best friend. I finally read the note and it reads a gift for you my old friend.

I assume it was from myself in the future, I wonder why I'd take the risk, unless if this dog is more important than I realise now. I take some money out and give it to the cashier, she tries to refuse but I insist as it's the least I can do, they then thank me, and I leave with the little dog in toe, surprisingly comfortable in my company, like it has met me before, maybe dog's are far more perceptive than humans after all.

As I reach my house and unlock the door he goes running inside, as if looking for something, he pulls out of a doggy basket from a cupboard near my desk I didn't even know I had, then paws on a nearby cupboard in the kitchen, much to my amazement I'm stocked up with months of dog food, these cupboards were empty before I swear, unless If I'm a sleep walker that likes to hoard dog food, I don't know what's going on, unless if my future self thought to stock up, as me and my companion would be together for a while.

As I look down at him he retrieves a doggy bowl from a cupboard dragging it towards me, I struggle to open a bag of doggy treats, so I take a pair of scissors, and carefully open the flap at the top of the packaging, he wags his tail excitedly as the biscuits flow into the bowl. I run the tap for a few seconds, almost splashing it all over the worktop, as the water fills the small bowl. I raise it out of the basin, almost sending it flying all over me, as he scratches up my leg for attention.

I place it carefully beside his doggy basket, and he runs over feverishly lapping up the fresh water, like he hasn't had a drink in months, he must have been starving in there poor thing.

I stroke his head carefully, poor doggy I see why he was left for me now, we're both alone in this world, and we only have each other for company...

Chapter 13 – Desolation Of Time

The wind seems to grow louder everyday as if it's calling out to me, I look out the window to see thestars as if gathered in procession. I wonder what it must be like to explore the expanse of space, maybe one day I tell myself, first I need to discover a means to stop this alien race from conquering earth once and for all, otherwise everything mankind has worked for, for 1000's of years will be for nothing, as we will be enslaved like mindless drones, doomed to follow our master's bidding.

It was only a matter of time before they discovered what I was trying to do, and they would come after me, but I would be prepared, the time bomb is only the first step in a multiple levelled plan I have for years to come, that's the advantage of being able to move freely through time, but everything comes at a cost, my health is merely a detriment to me at this time, as I grow weaker with every passing day.

I need to obtain a blood sample soon, to cultivate a healing salve from a combination of their blood, and the decaying tissue which seems to be mutating my skin cells like cancer attacking my flesh, it's only a matter of time, before my organs become affected as well, and how long will I have to live then, days, months, or possibly even years? I can't even see a doctor as they would think I was mad, and would lock me up like a lunatic, raving on about time travel, they would call the white suits before I could blink.

I needed to discover this on my own, as I look down at my dog, I stroke through his fur, and tell him reassuringly, it will be okay boy, he can tell something is wrong with me, as he sees the light fading from my eyes with each passing day, as I slowly get drawn into an everlasting sleep, I can feel the equal push and pull of time itself, my mind drawn into a void between timelines, as if it's trying to tell me something of my future.

I'm not sure where I'm going to go next, I need to complete some repairs on the machine, but it should be ready for another trip in the next three days, until then I will make notes of my progression. I feel challenged like never before, as a human and also as man. I feel like I'm breaking a barrier between what is known, and what is very much unknown to us.

I feel as my eyes grow heavy in the piercing moonlight, I look up once again, and see the full moon, as if it is staring directly at me. I implore it for answers, but I find no conclusion to my questions. I feel as if I'm past recognition of my timeline, as it seems to be out of alignment with the world, I'm not sure when or where I am now.

The piercing gaze of the moon cuts through me, like a beam of light cutting through solid oak, a tree falls silently in the forest of time, lost in the fires created by the race, who would seek to enslave us, or destroy us all if necessary, in order to progress themselves, sacrificing our free will in the process of assimilation into their collective hive mind, of sentient goo.

This reminds me of the old terminator movies all over again, it's a shame it's not as simple as immersing them in molten iron, then it's over, no...this story has only just begun, and it will be years, if not centuries in the making before the final fruit will fall, and their race shall be plucked from existence, once and for all, as this is not a case of one man, it is the case of all human kind, being able to live and breath, as human beings with free will...

Chapter 14 - The Second Sun

12 years later

Quit your yapping dog I'm trying to work...ruff ruff! I know you're hungry, but...hey wait a second okay that's the phone...hello who's there? all I could hear was static on the line, I repeated myself who's there please? WE...KNOW...WHO...YOU...ARE...I dropped the phone in shock, as I said quivering...who are you exactly? WE...ARE...THE...END...OF...YOUR...RACE.

I slammed the phone down, and took some pliers to the hard line, I had to move fast, I took the dog to the machine, as he had to make an escape. They were getting close now, I could hear the thunder outside, I knew that all too well, from when the drones came, another few seconds and I would have been cooked alive, I only had time for a small jump, 12 minutes in the past should suffice.

The quantum shifting is something you never get used to, as I quickly pack up my things, I look at my swiss watch, 8 mins before they attack, I set up the quantum manipulator in the centre of the room, 10 mins should be enough time to make my escape.

I sent myself to the coordinates directly east from here on the hillside looking over at the house, at this precise time, with a 30 second delay, to account for spacial interference. The familiar hum as the machine wurrs to life, I land on the hillside dog in hand. I can see them landing just near the house, patrolling the outside before moving inside.

A normal human being wouldn't be able to see this but the device works in a way similarly to quantum shifting. It slows down time to the smallest increment, then rapidly accelerates the atomic structure of the space around it, causing a massive explosion, but since it is contained in a quantum field, only organic matter in that field is destroyed. I can feel the surge as a wave of energy passes through me, and the aliens are atomised. I can now scavenge parts for my machine, a job well done!

Chapter 15 - The Quantum Professor

I was so wrong about the side effects of time travel, as it's slowly killing me, I can feel my internal organs, under the weight of time itself, I need to find a way to extract the alien DNA before it's too late, but how...I raid an abandoned hospital in the future, post invasion and grab some painkillers from a locked cabinet.

I try to make a quick retreat but I hear someone wheezing behind me 'please...help me...' I turn around and took at them, gas mask firmly affixed to my face, and I walk slowly towards them, and kneel down next to them. The woman is in her late 20's but her body looks destroyed by fever of some description, she tried getting up but her skin looks dry as a fig, and her hair is wiry like that of an old woman, but her eyes are still that of a young adult discovering the world. She says to me 'the pain...end the pain please...'

I tell her I can't but I rummage through my bag, to see if I can leave her something to help ease her suffering, I realise that morphine is basically induced suicide for such a high risk adult, but I have no choice, I give her the vial, and a syringe and tell her how to use it she says 'thank you...wheezing to herself...take this.' She gives me a small handgun she carried around on her person in her trouser pocket, I did not even realise she had one, this might come in handy, as I need a dispersal method for this strange alien metal that I've been researching, so I might be able to disassemble this for something else.

I say 'thank you, and good luck' as I make haste to leave, I can hear the life slowly go from her body, as there is a long sigh, and I look over my shoulder to see her lifeless body slumped on the floor, I try to force back the tears as I must carry on.

A scout ship was scanning the area as I passed by a nearby window and peered through, I had to make haste, dam it! I saw a drone hovering down the ward, so I quickly hid under one of the beds, I could hear the familiar whirring as it hovered across the ward, before coming across the young woman, then saying life signs minimal, prepare for extraction.

It then scanned her from top to bottom, then somehow materialised her into the little sphere, then it made to leave, firing a pin point laser shattering the window, I looked through the glass, and saw it hover down to the scout ship below, what where they going to do with her exactly, if they harmed her I would never forgive myself, but I know what I must do, I must stop this once and for all, so I couldn't let her sacrifice be in vain.

The machine was only a few yards away when I heard a loud whirring sound behind, little did I know that drone came back inside for a second luck, as I slowly looked over my shoulder. I bolted for the office, where I hidden the machine, luckily I fitted geometric computer to calculate my position spatially in the world, so I could travel between different points miles apart, otherwise I'd have no chance.

As I activate the remote activation sequence by shouting 0012 alpha it whirs to life, and homes in on my position. As a laser makes contact with my shin it whirs to life, whisking me away to the time before the invasion once again. I bandage up the wound to my ankle, it's only a very minor burn mark, but it did leave strange cartographic markings on my shin, similar to the markings left by a brand, most unusual.

I realised I needed to start work immediately on a smaller model, as this was far too big, and slow to activate to keep bringing with me everywhere, so I hard to start work on a time watch immediately, it's simply a natural progression...

Chapter 16 - Time's Source

As I travel across different timelines I start to notice my own particles shifting consciously in my body, as I soon feel myself pulled forcefully by an immeasurable power, like air being forced through a windsock, as my body convulses, and is forcibly ejected on the other side...I wake up face down in a completely white space, as I attempt to get up, the ground ahead of me begins to shine leading the way towards the centre, towards some kind of complex ahead of me.

I feel like I'm walking into the heart of a star as the light becomes brighter as I draw closer towards the strange pallid structure. I walked through the doorway, and all that I see before is a strange psionic eye, darting from side to side. I then hear a voice in my mind, welcome to the axis of time mortal, you are the first to visit this place for several thousands of years, we speak to you in your language of choice, as I know the voice of which many millions of civilizations speak.

I attempt to speak but then it says in my mind, don't speak it think it, we shall hear you in your mind much more clearly. What are you exactly? We are existence, we are the stasis of time.

We were created by the one's you seek to destroy, but we..know..this..must..be..done..otherwise..wrrr.it starts to hiss and combust, as the eye spins rapidly before falling to the ground, it then says with it's dying words...you...must...stop...them...all...it then disappears in an immense flash of light.

There is a console ahead of me in strange text that I cannot read but the watch in my pocket suddenly begins to react violently to this, as I hold it up in front of it, a proximity sensor comes up on the watch, like a flashing beacon, which seems to grow louder as I move it towards the console, as I place it on top of the screen.

It whirs for a second, before coming up with a series of numbers, and then scrolling through the alphabet, before the screen suddenly goes blank, and the words...WE NEED YOUR HELP...come up in large writing on the screen. I step back almost falling over on a small step in the centre of the room, I need to sit down for a second.

I then notice it changes again and flashes some coordinates and a date/ time, I attempt to input the coordinates on the time watch, and it says coordinates unknown, choose different data.

The console then changes as the screen fades away, disappearing leaving only an empty void. I look into the hole that's left, and I see a familiar device, it's the quantum watch, but a newer version, maybe I won't have to reverse engineer this one after all.

I place the old one back in my pocket, and then lift up the device from the cavity inside the machine, and attempt to wrap it around my wrist, it then clamps down hard, and two small prongs go sharply into my arm, I recoil in agony, as it says DNA confirmed, and then whirs to life, time til jump 3 seconds, WHAT! I say loudly...2...I don't even have time to react before...1...I get shifted to a new timeline...0...and then I was gone...

Chapter 17 - Raptors Claws

3000 BC date unknown

I awake in a prehistoric looking area, either I fell unconscious from the jump, or I was teleported back to the time of the dinosaurs, as a t-rex approaches me, I close my eyes holding my eyes in front of me, but then it seems to walk off disinterested. I then realise the watch says CLOAK ACTIVE, in large writing, so I can walk freely without coming to any harm, unless if something steps on me like an ant under a human's foot...

I walk around a bit more trying to find my bearings, the watch signals some kind of beacon ahead...a beacon in this timeline, but how? It could be something the alien's left here in the past, maybe it's some tech I can use.

 I eventually find it after navigating through the trees, some prehistoric creature is prodding at it, but as I lower my cloak it hisses at me, before I draw my sword, and it runs away in fright. I clean off the slime on some nearby leaves, and attempt to examine the strange device, maybe it's some sort of tracking pod from a crashed ship, similar to a black box, I need to do a bit more investigation, as this metal is hard to come by...

I re-enable the cloak and make my way through the trees, to hear a faint buzzing sound, as I cover my mouth, the smell of a decaying alien corpse fills my nostrils. It is surrounded by small aphid like creatures, luckily nothing can see me as I climb up the outside of the pod, as I need to take a sample of it's blood for analysis.

I look around to see if there's anything else I can retrieve, whilst trying not to wretch from the repulsive smell, but then as I make to leave, I realise that I must have accidentally disabled my cloak, I hear a familiar sound like that of Jurassic park.

As I turn around a feathered creature, about 4 foot high is sniffing around me, and signals to his friends. I then realise I have alerted the raptors to my presence, so now is the time to run...As I quickly try to activate the watch it can't get a signal through all this foliage. I need to find an open space, so I make my way back to where I shifted here.

I eventually find the open area which I came from, but I can't establish a signal, before I know it I'm surrounded, then something miraculous happens...the t-rex comes back and scares away the raptors. I didn't realise this earlier but it was protecting it's young, it didn't mean me any harm, as I make to leave, I nod thank you...

Chapter 18 - The Inverse

I find myself in a place where everything is parallel to what I expected, up is down, left is right etc. I'm not quite short what I'm doing here, as the sky is a shade of red, like some dark mirror, looking back at me. I try to jump and I realise gravity is still present, albeit that I'm jumping down instead of up. I then notice there are glowing pads ahead of me leading down, or is it up into the red sky. As I move closer to them I can feel gravity returning to normal, as I am pulled upside down towards them, it seems like they are meant to guide me to my true destination, as the ground above me, looks extremely hostile, like the surface of Mars on earth.

I make my way down or is it up? I wonder as I ascend the platforms ahead of me. I find a glowing compass at the end, I wonder what the significance of it is, as it sits atop a pedestal omnipresent here, like it holds some meaning. I then realise the prism is glowing rapidly, as I move it closer to the compass, it absorbs it until itself completely, I then realise what it's for. It's to show me the destination in the current timeline, as is predetermined by the events which have already occurred in the future, which is extremely useful in my current situation, as I have no sodding clue where I'm supposed to be going, as everything is upside-down!

I look around and realise the platforms behind me are too high, for me to jump back up to, so I make my way towards a tilted platform on the other side just above me, and climb up the edge of one side of it. I do the same three or more times, before leaping to the ground above me, grabbing the shallow earth below my feet as gravity corrects itself, pulling me towards the ground in.

I reorientate myself, and try to get my bearings I look ahead and see nothing but sand in the distance, but I continue onwards just encase there is anything else to explore. I walk for what feels like forever only to find nothing, I wonder if there is any further purpose for me being here, but then I notice there is a light in the distance. As I walk closer towards it, it's almost as if it can't be seen from certain angles, like the light is only travelling solely from one point in space, which is impossible unless if it was somehow bending the space around it.

I draw out the prism to try and get my bearings, and it glows brightly in the direction of the light so I continue onwards. I feel like I'm making no progress as I walk towards the light, so I wonder something...perhaps if I walk backwards maybe it will move closer towards in a similar way to the platforms from earlier, as gravity is a bit strange here.

I try it and sure enough it has some effect, just not the desired one, as I move backwards it grows larger and more distinct, almost as if it's going to engulf the whole horizon, then there is a bright flash, and where I am now I do not know for certain...

Chapter 19 - The Forgotten Realm

I find myself in a place between time and space, there are lost spirits crowding all around me, was this what I saw in my dream I wonder? The prism glows excitedly as I draw it from my pocket, it creates a path for me through the fog, as the spirits part to show my way. I look around, and they all look so disenchanted, so formless...like grey ghouls hovering in the darkness. I feel incredibly sad all of a sudden, but I must carry onwards. I eventually find my way to a strange alter, and the prism suddenly focuses it's light towards my sword, so I place the sword on the alter, and something strange happens...

A violent and incredibly powerful burst of light, fills this ghastly hollows, the moans of the dead, can be heard resonating with my very soul. The process is nearly complete, but the ghosts grow restless, moving towards the light, I cry out NOOOO! but it's too late, their souls are extinguished in the brilliant splendour...but there is still hope.

They do not disappear completely, they are replaced by brilliant beacons of white light, like angels sending their way into heaven. The process is almost complete, but then I hear a mighty roar as if from hell itself, what kind of creature could make such a noise?

A horned monstrosity comes barrelling towards me, it's legs wrapped in obsidian chains as it crawls along the ground, using it's incredible strength to it's advantage. As I tentatively close my eyes, it stops letting out a loud roar, but It cannot come near the light for some reason, like it's protecting me...maybe that's why I'm here, so that I can slay worse foes than the grey.

I retrieve the blade from the pedestal, and hold it high in the air, to drive the horned beast back. It cowers back in fright, as he crawls away I realise it's only a matter of time, before he comes back again to torment these poor souls, so I raise the blade, and without even realising it. I strike down creating a cascading beam of light, which strikes directly through the demon, sending it's collection of souls, back to hell from whence it came.

The souls now free of it's grasp, begin to take form, women....children, as well as men of all shapes and sizes. They all look over as if to thank me without words, they are finally free. As I make my way to leave, I see them all waving, into the infinite expanse beyond the light...

Chapter 20 P art 1 - The Invention

13 years ago I came up with a theory about quantum relativity, and how it could reverse the effects of global warming, or at least help prevent further damage, by creating a sustainable energy source. It would utilise the flow of energy through time and space, from a place known as the zero dimension. My theory was abandoned as nobody took me seriously, and I was almost kicked out of university, for my radical theories on quantum relativity, and hive mind electronics e.g. borg tech.

I wasn't the only person however who believed in this theory, my class professor also my mentor, A. Stanton believed my theories could be possible, but it would take year, for current research to progress that far. He recommended I apply for a research grant at Yale, and try to get military funding, from their applied research division.

It was a little known branch of the military, which specialised in future proofing today's military, by allowing them to be always one step ahead of enemy forces. If I could prove that my technology, had advanced military applications, such as the Philadelphia experiment, they might grant me the funding to build a prototype.

It took around 53 days for me to complete my masters after the conversation with my professor, after that I was free to do as I choose, so I headed straight to Yale's ARD. This functioned in tandem with the military's applied research division, much like a right hand inside the educational establishment of Yale.

Professor Stanton, told me he knew someone who worked there, who would be willing to listen, to what I had to say, his name was Harry Seymour MD, and was a specialist in particle physics.

He had a keen interest into the implementation of quantum computing, and the learning computer model. I wanted to approach him about a theoretical program, which could simulate the workings of my 'machine' without any risk to myself or others.

We worked for days, programming a basic data structure, with different computations for mass, and body type, accounting for energy output, reaction delay, and also gravitational inertia, especially when transposing large amounts of energy. We 3D printed a basic model about 5 inches tall, and 12 inches long, which was the basis, for the basic structure of the machine.

It wasn't long before the military got wind of our designs, and decided to step in with a big pay cheque, if we continued our research behind closed doors. They had a team of over 50 people waiting for us, but I told them straight away we only needed five people in the room, otherwise it would be too crowded for me, and my colleague to think clearly.

They thought it preposterous, but they respected my wishes. 12 days later we had built a basis chassis for the machine, using an old 4x4 engine, which was lowered into the chassis. The fuel source was a mixture of hydrogen oxide, to allow for a massive burn, to get the quantum accelerator going, with diesel powering the engine itself.

The machine was raised about two feet off the ground, and the hydraulics were connected to a sole wheel on the back of the machine, to simulate friction, whilst accounting for gravity, whilst remaining immobile. I turned on the engine and heard the familiar whirr of the machinery, within the vehicle if you could call it that, it was basically an engine, with a wheel attached to a metal frame.

The catatonic motion of the machine kept going for another 30 days or so, before we took away the breeze blocks under the machine, using a forklift to raise the pallet under the machine, and some local construction workers, who were given a nice pay-cheque to keep quiet, and help us raise the machine up a few inches.

We pulled out the breeze blocks, and lower it onto an anti-friction mat, which we purchased quite reasonably. We thanked them, and they left to do their jobs once again, unwittingly helping us achieve, one of the greatest scientific breakthroughs of our lifetime. The engine was replaced with a metal fuselage, which was a converted cathode tube, so we could do basic ignition tests, to see how much power output, could be produced on a short burn, at different output levels.

We eventually managed to manufacture, a more sustainable fuel source, which their scientists had already been working on before we arrived. A dutirium reactor, the smallest of it's kind, it was partially powered by hydrogen fuel, and dutirium along with fans, which ran along an exhaust system, which turned small turbines, creating kinetic energy with which to create sustainable energy for the machine, similar to a nuclear reactor.

The heat was created using plasma lasers at a specific frequency, which were powered by a series of dilithium capacitors, along the outside of the tank. They were carefully placed so as not to create static, between the moving parts inside the machine.

These were used to store the excess energy, created by the machine for future use. We fitted a large coolant tank to the top of the engine column, with the tanks placed underneath it, it would release water into the tank creating steam, which would turn the turbines, creating energy. It was extremely heavy however, requiring multiple people to lift it, as it needed to withstand, a huge amount of stress.

Chapter 20 Part 2 – Quantum Necrosis

As it was also the only means to stop the machine overheating, as well as allowing it to create sustainable energy. The plasma laser was also designed, to create a gas sink for the hydrogen atoms, to force nuclear fusion within the small fuselage at the bottom of the chamber.

The excess water trapped within the chamber would be vented off, and burned off using a process of nuclei synthesis, which occurred within a large basin, at the very bottom of the trunk. This explosive reaction was contained within an approximately 5 inches thick tank, crafted from the same metal as space x rockets.

This was designed to trap the gas inside, allowing the machine to work The excess vapour would be released via a series of pipes vented from the machine through the base of the frame, dispersing a small puddle on the ground.

The time travel mechanism as it were, consisted of three dials. The first of them was an atomic clock, another one was a depth meter, showing the depth above sea level, and another dial showing the current present time, which was a black quartz clock, specially made to withstand depths of up to 1500 meters below sea level.

The front casing of the machine, was made from a hard shell of meteoric metals, in a similar process to how diamonds are formed, through a process of super-heating, and compression, then they were formed around large moulds, using huge industrial machinery.

This formed the front of the machine, so that it would be able to withstand the incredible pressure, of interspacial travel. There was only space for one occupant, accompanied by a flight computer, which was a small screen in front of the passenger, similar to a modern GPS. It would tell them their current location, and allow them to communicate across time, theoretically speaking of course, as we haven't been able to test that functionality yet.

The final piece was the 3 inch think glass pane which covered roughly half of the machine, and was designed to open on the left side, for safety reasons. This was to prevent loss of pressure, and protect the occupant seated in the right side of the machine.

It allowed for full 360 vision, around the machine at all times, and was designed using a similar material to gorilla glass, except it was superheated and shaped, within a cast iron mould, in a similar way to the front of the machine. The glass could withstand anything from a speeding bullet, to a meteor strike, or even possibly a nuke.

There was also a gas mask just under the chair, which had access to an emergency oxygen supply, encase the machine locked up, as the only way out was through the unbreakable glass dome, or the left side door. The right side was completely sealed, as if you're going to fall down a crevasse during entry into a timeline, you're most likely going to fall towards the right, since most people are right handed. They thought it would be more sensible to seal that side for safety, so as to prevent any incidents.

The door was only operable, using a small device similar to a watch, powered by a solar battery, which was keyed to the DNA of it's user, using a series of pins which fed directly into the machine. This was to prevent theft of the machine, and would self destruct if stolen from the wearer.

This would cause it to return to it's destination, instead of going nuclear. This would then fry the internal computer circuitry, so it couldn't return to that timeline. It would then notify the lab, of the fracture in the timeline, and to destroy immediately in a safe and secure manner.

We soon began testing the device, but it caused some unwanted side effects...we recorded in our notes about a sickness which affected around 90% of subjects, which we called Quantum Necrosis or time sickness. Their translocation signature also changed causing them to convulse violently, with strange echoes through time, causing their quantum signature to fracture, even after only being shifted a few minutes forward in time.

My notes were inconclusive as to any further side effects, as human testing was stopped after the tenth subject died from violent convulsions, causing damage to their temporal lobe, which led to their inevitable death. The only subjects that seem unaffected are those with a naturally occurring protein in their blood, which we called protein Z.

We were both unaffected by it somehow, further tests revealed we did in fact have the protein however, but it seems extremely rare, but we're not sure why it's in our genes. It could that during the creation process, somewhere along the line, certain individuals were given preferential genes.
A large unit of time, seemed to be lost in their memory, in a similar way to dementia patients, they couldn't recall where they'd been for the last 5 or 6 hours after shifting, even for us the time between jumps would become lost of confused, like we had just appeared in a new destination, after just getting back to the laboratory.

We made notes of this time lapse, and estimated the amount of time that had passed, written it down then checked in against the present time. It wasn't as perceptible for us, but there was around 8~10 minutes missing from our memory, after our mass was shifted through time, so recording this phenomenon was very difficult.

We felt like this was just a part of the progression of the technology, and we just had to manage our time carefully, to prevent more severe effects, such as neural degradation, or possible seizures in the future, or possibly even death from overexposure.

Chapter 21 - Saviour Of The Past

I never imagined myself being referred to as a hero, but as I return to the future, to see the world burning, I wonder...why does it have to be like this, why can't we coexist in peace? I then see a giant mother ship descend, not as cleanly built as I might have imagined, it looked more like scrap metal, just fused onto the external fuselage of an existing structure, like a smaller ship, built up to appear much more sinister, or maybe just more deadly. It's laser reticule then focused down on the city below, firing a massive laser, killing 1000's of people within seconds, their thundering cries could be heard for miles, I couldn't take this anymore. I headed back to the time just after I first shifted. I saw myself standing their working on the machine, unknown to me my doppelgänger was standing within mere feet of me.

I hide quickly behind a wall, as he heads in my rough direction, I forgot I had to take a leak at this point, as I make my way quickly to the machine in the present time, and tap the coordinates for Pompeii. As he comes back, unbeknownst to him, he heads over to the machine, without even realising someone has changed the time on the machine. I can feel myself ageing as my voice becomes course, and my beard becomes more grey by the second, I've just extended my timeline by another three years, but it must be done...

I then head back to the school fair, looking like an old man, with a long straggly grey beard. I tell a young boy exactly what he needs to hear, I then leave a crumpled note for myself, roughly where I stood in the hallway so long ago. I then recite the lines to myself, exactly as I remembered hearing them so long ago, I look at the other piece of paper I crumpled in my pocket 12:45.

It was almost time...he would be here soon, I needed to seem like I knew a lot about the future, when in fact less than 10 years had passed. The signs of old age, were merely a side effect, of too much shifting through time, but I needed to act like the wise old man which I appeared to be, like in the story books I once read, brushing the hair from my face with my hand.

He rushed into the canteen and I told him, don't come near me! We cannot occupy the same space at the same time, he looked hesitant, so I lied through my teeth and told him. Here take this I have several of these, knowing full well I only have two, the one from the axis of time, and the one I gave myself from the future, the one I'm giving him now. I haven't even created yet in this timeline, but time is a funny thing, and it always has a way of working out like that. I then disappear shifting to another timeline.

I then look over at the the figure in the water, as I stand atop the roof of one of the ruined buildings in Pompeii, I survey the area for any signs of life, but there are none, dam! I tell myself, I could have done something, but I can't, this was meant to happen this way, there can't be any witnesses.

I then get the strangest sensation I'm being watched. I then see a drone behind me, and I slash through it with my blade, luckily he didn't notice my spectacle, as he's still staring at the strange alien machine on the horizon. It's at this point I realised I must make one final trip, and this time I don't know where, the civilization which the prism calls home, Atlantis...

Chapter 22 - The Metronome Effect

I find myself in the middle of a desert, and the prism reacts violently to something ahead of me. As I take it out of my pocket to find out what's going on with it, a sharp laser fires into the distance opening a doorway, seemingly into another dimension. I walk towards it wondering what lies ahead...I find myself in an inverse space, where up is down, and down is up.

There are stairs everywhere, and doors leading to multiple places. I hear a noise behind me like a clock, but not...I then realise there is a giant rock metronome, with a metal spindle suspended in the back of the structure, each strike of the metronome, shakes the structure violently, as I walk down or up?

I'm not sure, I find myself entering another doorway, and I emerge on the other side of the room. I look up to see where I started from, as I put my hand through the door, my arm appears above me making me feel bad vertigo all of a sudden. As I enter another door, I'm walking along the side of the wall, towards another door, where will this end I wonder?

I enter the last door, and find myself in a large room, which looks like the inside of a pyramid, with sloped walls and colossal statue sitting atop a giant throne, about thirty feet high. I see a strange orb floating ahead of me, and the orb is absorbed into the prism enhancing it's power even further. This is when the metronome suddenly comes to a stop, and the room shakes violently, as sand pours from the statue it seems to come to life. The entrance twists away into a corridor behind me, the walls become steps circling round the outside of the room.

As I attempt to break into a run, I can hear the giant colossus, smashing against the walls to pull himself up, he then roars shaking the entire structure once again, and smashes the metronome back into action, which causes gravity to become extremely heavy, but the prism seems to react, creating an anti-gravity bubble around me, using it's newly found power.

I struggle towards the exit, but somehow through pure force of will, I feel myself moving closer, just as the colossus' hand goes to grab me, I feel as the air is sucked out from underneath me. As he clamps down hard just inches from the back of my head, I launch myself through the portal back into the desert.

I make my way back towards the quantum translocation point, and I hear something trying to force it's way through the door, as a giant hand forces it's way through, attempting to slam down on me.

It tries to pull itself through the opening, as the prism begins to glow, I place it between my hands, aiming it towards the portal. As I force the door shut, the colossus screams in agony, shaking the sand beneath my feet, as it clamps down around it's arm, cleaving his arm off completely, it then disintegrates back into the sand, as the colossus disappears completely.

Chapter 23 - Advance tracks

I save up all my energy for this jump, as I seemingly find myself a very long way from home, I will recount how I got here momentarily...it all started about 5 days ago...I was somehow stuck in that neverending expanse. The quantum manipulator wasn't working, so I needed to find a power source, so I wandered, until I found an oasis. I stopped for a few hours to think, then I realised I had the prism, and it showed me the way, I was standing a cliffside, wondering where to go next, little did I know that it could potentially also be leading to my death...

That golem was only one of many guardians of these ancient ruins, we weren't actually in a place as such, we were outside of normal space, I reached here thanks to the prism, but above me...is millions of gallons of water.

The GPS tells me, I am deep in the marianas trench, deep under the ocean, inside a giant pressure cushion created using Atlantean technology, to trap whatever 'it' is that caused Atlantis to sink in the first place, perhaps it didn't want to be found after all...

The desolation of this place, begins to get to me, sending shivers down my spine, I look around for a way across. I see a tree teetering on the edge of the chasm, I attempt to push it over, with a couple of thrusts with my shoulder, it falls with a creak, then a loud thud, as the sand shifts underneath it.

I shield my eyes from the cloud of sand, as I attempt to crawl my way across, but it snaps under my feet. As I leap to grab the ledge, and manage to somehow pull myself up, I see it bounce off of some sort of invisible platforms, about 100 yards below. How do I get to them I wonder? I ponder on this for a moment, before coming to the realisation...why the hell is there so much sand here, what could it be hiding?

I attempt to dig a hole near the edge of the cliff, but much to my surprise, I don't find bedrock underneath no...I find thick metal slabs, intentionally and thoughtfully placed along the edge of the cliff, like a building of some description.

I keep walking along the edge until the prism reacts to something, and begins glowing brighter as I keep moving forward. I then see something flashing underneath the sand, a voice then asks for identification, the prism is able to convert the sound waves, into a language I can understand, it then somehow hacks the device, and opens a stairwell leading to an lab below.

Chapter 24 – Explosive theory

I journey further into the depths of Atlantis, and find a small security console flashing in an unknown language. It's somehow translated by the prism, as words appear on screen in English, but it then appears as; ERROR! ENVIRONMENT CONTROLS NOT RESPONDING! It keeps repeating this as a loud siren, is blaring ahead of me, almost deafeningly. I raise the prism in the air, and attempt to scan for a power source, and a panel begins to glow in the right side wall, which flashes green. I walk towards it, and press my hand against it, and it opens up to reveal some kind of power relay.

It looks like one of the power conduits has exploded, but I have nothing to solder it back together. The prism glows once again, and then I realise I have a laser beam in my hands. I move a couple of feet back and aim the prism at the power relay, it then fuses the wires back together, and power is restored to the facility. The monitor turns back to green, and the lock-down is disabled.

I attempt to find some form of life, but after 100's of years, I doubt anything or anyone survived here. I need to find a power source to get home, as I believe I see a shadow standing in an adjacent corridor. As I shout; Who's there? It quickly runs away in the opposite direction, as I give chase the door slams shut behind me. I eventually find the creature, cowering in the coroner of the room, as it says please! don't hurt me!

I didn't mean to kill them! It's one of those aliens, the grey creatures, but it's talking English, how?...What is this place? WHY ARE YOU HERE? I say pointing my blade at it's neck, I'm a researcher from Atlantis...those creatures...they were studying us, they wanted us to build a machine, to enslave the human race, use them as...vessels, we refused and they cannibalised us.

We were almost wiped out, about 263 years ago, hence the advanced shields. There powered by the strong water pressure here, they kept me alive, but turned me into this...thing, this abomination! I don't want to die, but I don't want to live like this either, I just want to be free of this pain!

What...what's wrong with your face? I touch my face and the black goo has spread, dam it! I still need to create a vaccine, it's spreading across my face...He hands me a syringe, and states here....take this, it's a sample of my blood. I used it for testing as somehow I survived the process, unlike my friends, who were drove mad, and died in horrific pain, I somehow remain lucid, and the alien antibodies don't effect me somehow.

The alien DNA seems to have incredible healing properties, I had to engineer a means to pierce my own flesh, using one of the daggers they left behind, but I didn't want to bleed out, so I chipped a piece off, and hollowed it out to create this syringe, using a high penetration laser, which was used for mining, and shaping precious metals, to be used as weapons or for tools. It might be a small consolation, but it could definitely save your life, if you're suffering from what I think you are.

You've seen this before, he replies yes...once before. We sent a subject forward in time, to study the evolution of humanity, they were gone too long however, and when they got back, their body was covered in black pustules, which burned their skin, like their skin had been set alight, and they writhed in pain, before dyeing in agony about 12 days later, hissing and screaming.

Their bodily functions shut down one by one, and we could not find a cure, for the viral necrosis, but you...you seem to be more resistant to it than they were. I have tried killing myself several times, but this body...this thing won't let me die, the other's left me here, hoping I would either die, or go insane trying to kill myself, as punishment for not creating their ultimate weapon.

I'm sorry to hear that, but I have something to ask you before I leave, do you know of a power source around here, I need to get back to the future. There is a generator powered by hidden panels on the surface, which use endothermic energy from the sun, using satellites around our planet far beyond the capabilities of modern technology.

As the small amount of research we had gotten from the aliens, allowed us to complex series of lasers concentrated the sun's energy powering our machines, by heating the solar panels situated around the planet. That is which gave us the power to enhance our scanning capabilities, and eventually led to incident 23...What is incident 23? I can't tell you now, but soon we must find the energy source you seek, this way...it's down the next corridor on the left...

Chapter 25 - Incident 23

What were you studying down here? We were studying genetic research, and the hybridisation of humans, with other creatures to create a sustainable outcome to our races survival. That sounds dangerous? It was in reality...most of the test subjects died days after being injected with the compound Rx2, except one...he is probably still locked in containment actually.

We called him warlock, as he had strange glowing purple eyes, and purple energy flowing from his hands, the mutation caused him to emit radiation from his body. He was created when we forced him to ingest a rare bacteria, which we found reacted strongly to human DNA, as tested on the deceased initially.

We believed this is what the aliens were doing with us, trying to project themselves onto our DNA. It brought them back in a mindless vegetative state for a few minutes before dyeing again. He volunteered for the project...his name was Jonas.

We called it incident 23 as he was the only subject to survive the experiment, oh shh....what is it? He's gone, he's somehow blasted his way out of the cage, we need to run, NOW! I can hear heavy breathing in the corridor ahead, and a voice speaks through my mind, I..will...find...youuuuu! An arm then reaches through the wall, as he almost grabs hold of my hair, he somehow phases through the metal, as energy courses through his body, he comes to charge at us again.

He looks akin to a wolf, his hair overgrown, with torn musculature, and strange growths on his arms and legs, pulsating like purple ooze flowing through his body. The grey figure then says quickly tapping my arm.

This way, we can seal him inside...we run towards another room, and he activates the flashing security console I saw before. He then goes to pull out the wires from the wall, as I say I just fixed that...the door slams shut on his foot, leaving purple ooze on the floor where his foot was, and dismembered toes, which seems to write of their accord.

The electrical field won't hold him for long, quickly this way. He says as he activates a panel revealing an upper level, with a ladder, leading to a ventilation shaft. We crawl through the tiny space, and eventually reach the generator room. He hands me some wires, and asks me which one's I need, I point at the standard 9 volt plug, and use a convertor to allow it to charge my device, at the correct voltage.

He then flicks a few buttons on the console and input mode changes to b, and it prompts me to insert the cable into the machine, it scans the input and retrofits an input socket for the plug. I then plug the other end, into the top of the device. He then says...he has found us, as he realises he appears at the door way. He then says quickly 'hand me your sword...you need to get home in one piece correct?...I will die here anyway, so let me help you save your world! '

I hand him the blade, and he slashes through the air, baiting the creature to strike, much to his surprise it does, as the device shows 88%...89%...I then says it's now or never, as I pull to power just before it hits 90%, grab the blade from his hand, and slash his head cleanly off, leaving it rolling across the floor, in a pool of purple goop. It's done I say to him, putting the blade back in it's holster, after cleaning the mess of the edge of the blade.

He then says...I'm...glad...the bite did a lot more damage than expected. The bacteria was attacking his already compromised immune system. As he says...go...go now...before it's too late for your people as well, the future is depending on you, my time has passed long ago. He takes a deep breath then continues, I hope you find a cure for the plague that ails you, and good luck on your travels through time...I input the coordinates and disappear, shedding a small tear, for the unlikely hero, that saved me in his final hours...

Chapter 26 - Quentin, and Schroedinger's Cat

Is it possible to be dead and alive at the same time? That is the question I ask myself, as I tighten the belt around my arm, and inject the strange substance into the brachial artery. I feel a rush of adrenaline as it hits my heart, causing me to convulse violently. As I start spitting, and pulling the table from side to side, the dog becomes restless, as it hears me crying out in pain. I then become still for a moment, but as I open my eyes, I catch my breath, and I can feel the strange substance, retracting the blood vessels in my face. It burns me to my core, but I can feel the acrid substance on my skin disappearing, as I remove the belt from my arm, and grab a nearby mirror, I can see the blackness on my face retreating from my flesh.

I go to stroke my dog, but he moves away from me in fear, I then realise what I have become. I have become like them, that's why he's scared of me, I then realise what I must do...I open the front door of my house, and say to the dog, go home boy, beckoning him through the door. As it cowers from me, I move away from the door, trying to hold back the tears, as I shout GO! ...pointing angrily at the door, whilst sputtering tears between my words. The dog scurries out the door, before looking back one last time, with a familiar ruff! As if to say...goodbye master...I then hear a rumbling above me, as the house shakes and a drone appears, comes down and steals him away before my eyes, time to test my new found strength...

I grab the blade from my waist, and step out in front of my house, exclaiming to them, come down and face me you cowards! 3 of the alien ships descend down, with their laser eyes focused on me, before creeping back in fear as they scan my DNA. I then fly into a rage, striking the blade into the ground, before charging towards the leftmost creature, swinging it up through the chassis of it's ship, causing fluid to rain down on top of me from inside the ship. As I run towards the left ship, and slice off two of it's three legs, sending it crashing down to the ground. The third tries to get away, so I grab one of it's legs, and pull myself up to the top of the ship.

I cut a hole in the hull of the ship, and physically throw the creature out from the vessel, as I get ready to board the larger vessel above...WHERE HAVE YOU TAKEN HIM! I shout as a cloaked figure goes running away with a tablet, and closes the door behind them. I look around for an override switch, and since I can't find one, I decide to make an opening of my own.

As I slice a circular hole in the door, and kick my way through to the other side, I can hear the familiar cries of my dog, as one of the creatures looks over and panics, accidentally killing him, with a shriek as my dogs neck is broken instantly. I scream NOOOOOOOOOOOOOOO! rage fills my entire being, as I feel myself grow in strength. The hybrid DNA floods my veins, I grab it's head, and physically crush it causing it to explode in a flash of grey goo. I then take control of the ship, and find the coordinates for their fleet, now they've made it personal...

Chapter 27 - Behind The Times

They attempt to close the docking bay for one of the larger ships, but I go rocketing towards it, practically destroying half the ships there, as the vessel careens to one side, colliding with the sides of the ship as I fly in. I manage to find some kind of weapons research lab, and I ask one of the labcoat aliens, where is your leader? They nod from side to side as if to say he's not here, I then say 'wrong answer!' I then send him flying towards the wall, as another one of them, locks me in again...why does this always happen to me...I then notice a strange device on a pedestal, some kind of magnetic launcher, with large silver balls in the chamber. I decide to try it on the door, It blasts a hole cleanly through it, and the lab operator who tried to run away. I might keep this, me likey! I say as I holster it in my left pocket, bulging out almost warningly, saving the last three rounds for later.

I then make my way down the corridor, but I can feel as the ship banks sharply, there trying to kill me before I can get to their leader, but that's not going to happen...not when I've gotten this far. I pull out the prism and will it to show me the way, it takes a few moments, but then it guides me ahead, then down a corridor to the right, and then left, and then the first door on the right is the control room.

I see another one of those foot soldiers, like the one earlier, manning the control console, but then it does something different from the others...It turns around in an instant. As if it could see me coming, and then it holds me up in the air, with some psychokinetic power, like nothing I have ever felt before. As I choke to catch my breath, it says to me directly in my mind, did you really think you could stop me so easily, you pathetic human, you abomination.

Your race are merely ants beneath our feet, if it wasn't for us, you would have never been created. I reply humanity finds a way...I mutter under my breath. What? he says...as he lets go of my throat. I take my chance, and plunge my blade into the control console, sending the ship careening towards the mountain side. As he says, NOOOOOOOO OUT NOW! WHAT HAVE YOU DONE? He then teleports away to safety, but then I realise I've trapped myself here, as I don't have enough power to shift in time...

Chapter 28 - Time's Epilogue

I managed to survive somehow, perhaps it was the alien blood, or sheer luck that kept me alive. The ship is just about shredded completely, but hopefully I can salvage some kind of shelter from it at least. It took me 3 days but I managed to build some sort of crude power source, from the parts that were damaged in the ships fuselage, luckily it had more than one generator to power a massive ship like this, which makes sense I suppose.

I just needed to work out how to get home, as I couldn't teleport that far with limited power. I try to make my way further up the mountain, making a basic cloak from the clothes, I salvaged from the ship. The same clothes which their scientists wear, probably just to protect their flesh from their experiments, rather than their modesty I imagine, since they don't seem to have any genitals, nor any discernible features, they look as if they are clones.

I also haven't seen any females, which is intriguing, maybe they don't need them as they're asexual? He was definitely the alpha of their race, he was much stronger and smarter than the other aliens I've seen.

I manage to get a stronger signal near the peak, bouncing it off a nearby satellite, and I manage to hijack the signal of a reconnaissance ship, sent to scour the wreckage.

I shift myself inside, killing the pilot of the vessel, and I have it turn around, whilst aiming the rail gun, at the scientist behind me, who was just trying to find out what's going on. I tell them not to move, otherwise I will shoot, and they seem to understand that, probably cause I have a weapon pointed at their head. I then park it on top of the mountain peak where I just was, and shove all of them out of the ship. I then leave, and take the ship back to my house, to try and salvage parts, so I can find the alpha of their race. This will hopefully allow me to create hopefully the first of many, quantum displacement vessels or time ships in the future...

Chapter 29 - Saviour Of Time

I need to find out where these aliens are converging from, but first I need to find out what that alpha was exactly, it was completely different from the others. They were like mindless drones in comparison to him, he had such immense power, like a thousand minds combined together, or maybe that's why they seemed mindless.

As he was controlling the entire force...If that's true why am I not affected? It could be because I didn't inject the alien blood directly, and I used the solution from that Atlantean gentleman who was transformed by it, but retained his cognitive behaviours. I wonder if that's the key to defeating them, mind over matter and all that, if only I can work out how these ships work...

I pull off the front of the control panel, and surprisingly it's similar to modern aeroplanes, they basically somehow have developed propulsion technology, way ahead of what we have.

This is despite everything else being very similar, except for the fact that they use a touch interface, for the navigation controls to make them much simpler, and it just means they have to rest their hands, facing slightly inwards to the left, in order to fly forward. The alien microbes seem to be able to activate the auto-pilot as if I was one of their own, which is incredibly useful for situations like this. I manage to fix the damaged parts of the ship, and I sort of understand how it works, I also try to find some time to fix the power source.

It appears to be a micro collider of some sort, with the collisions creating massively increasing bursts of energy, with the force being increased to 1000 kelvins, the amount of energy produced is off the scale, and it's probably what allows them to achieve space travel, in such a short time compared to us, to travel 100's of 1000's of years in our lifetime to do the same as them.

I still don't understand why they would risk coming back here, even though I know they say they created us, but why take the risk? It is just the fact alone, that they seem practically immortal unless...there race is dying, and needs to evolve somehow, so it could be that certain humans, can become host to a more advanced race of aliens, like the centuries old man, with the body of an alien from Atlantis.

I suppose anything could be possible given the current situation. It all seems to be up and running again, so I see if I can set any coordinates. The prism isn't giving me any feedback, so I guess I'm on my own from here on out...

Chapter 30 - The Emperor's Arrival

He is coming...he is coming...the emperor is coming...the voices said over and over in my head, the more time passed, the louder they become. I shouted, GET OUTTA MY HEAD! I sent equipment flying, as I flew into a rage, swiping at the air. My head was pounding like it was stuffed full of information, that I had never seen nor heard of before. It seems like the alien's knowledge was flowing into my mind, like a one way transit system, and I'm not sure how much longer my brain could cope with this...

It was over three hours before the pulsing of information began to halt, then one voice said...he is coming...he is here...the ground shook around me, as a massive ship began to descend outside my house. I took the back door, and decided to make a run for it, I knew I wouldn't make it but I had to at least try to escape.

I could see the lasers focused on the front of the house blasting it clean open, in an attempt to find me inside. My life's work would be destroyed, but I didn't care, what use would it be if I was dead, my work would be in vain, there would be nothing left, the world would die.

I slid down the hill into a nearby river and swam upstream, attempting to escape their viewfinders. I eventually made my way to town, and the voices became quieter. I went into the local bar and asked for a small glass of whisky at the bar.

The bartender remarked I didn't look so good, as I found some crumpled notes in the back of my pocket, I was saving for an emergency, now was as good a time as any.

I ran my fingers across my eyes, and I could see what he meant, my eyes were becoming extremely blurry. I felt the blood rushing to my face, I just said to him yeah...I haven't been sleeping much lately, it's too hot up here these days.

I pan my eyes across the room, trying to see who's around, as I know they can blend in with the surroundings. The bartender says 'yeah it is, I don't know what happened to the rain, but I don't mind.'

I then notice a strange man sitting in the corner reading a newspaper, his eyes have a glaze to them I can't quite understand. I notice he has ordered a drink but it's untouched, as he raises his head towards me, his eyes glow red, just for a moment, as if recognition has just crosses his mind.

He draws a small futuristic looking firearm, without even reacting I shift towards him, a few seconds before the shot, smashing his wrist against the small, and I silently take him down. I say to the bartender, who didn't see what happened, this man looks like he's had too much to drink, I'll take him home if that's alright.

He says sure, there's a hotel just on the corner, it takes in vagrants who can't make it home for the night, they pay in the morning. I can feel the weight of the creature bearing heavy on my shoulder, but I need to keep a low profile, I can't alert anyone to what I'm actually doing. I find the nearby inn which he spoke of, and the receptionist gives me a key to room 12, she directs me down the hall on the left, I say thank you. I can feel her eyes watching me as I walk down the hall, so I quickly shuffle into the room. I can hear a knock on the door a couple of hours later, she says in a low voice is everything okay?

yeah everything's fine, I just busted my eye on the door-frame, on the way in, as he jolted when I walked him into the room. I'm just cleaning myself up before I make tracks, I'll be out in a few don't worry miss! Okay sir if there's anything you need just ask, thanks.

I reply moving towards the door, and I listen for her footsteps, hesitantly moving back towards the body. The human form is starting to shift to something more alien, the camouflage seems to be wearing off, as I rummage through his pockets to find an ID.

There is nothing on the body, but he does have a small minallium blade, which I'll have. The lady comes back and lets herself in, saying sir I have a call for you, sir? I'm already long gone, and she is shocked to see the man's clothes, are all that is left, as I took the body away for an autopsy, as I want to study how these creatures tick, and how to kill an alpha of their species...

Chapter 31 - The Time Engine

As I dissect the alien corpse, I find a lot surprisingly similar to human beings, could it be that we were created the same way as them, is it just like he said? I won't know for sure until I meet the leader of their race, and find out for certain. I had to first finish work on the engine, using my old designs, I updated the machines frame, to make it lighter and more compact, using a frame from an alien ship I had discovered in the arctic.

The strange metal seemed to mould freely in my hands, given a bit of a push, it warped right around the front of the ship's machinery, as if some kind of nano machines constructed it. It intrigued me somewhat, but I couldn't find out what it consisted of without more advanced equipment, which I don't have.

I find a small device like some kind of transceiver, buried deep within the machine. I take it apart and manage to translocate a signal, to somewhere up in the stratosphere. I link it up to the machine's onboard computer, to see if I can find out any other information, such as it's relative speed and size.

It manages to hack into the ship's onboard computer, and reads ten lifeforms on board, but one of them is much larger than the rest. I try to hack into the binaural network, using my newly obtained alien DNA but to no avail. I can't seem to connect with the others, it's like I'm a walking receiver, but I can't communicate directly with them, dam! I tell myself...wondering what to do next, I replace the coolant tank with a much lighter more streamlined version made of minallium.

I replace the fuel tank with an upgraded version I managed to retrieve from the lab where I used to work, luckily nobody really looks around there much at night, so I was able to shift in and out in a few minutes. I then finally get around to compressing the dome to a smaller size by melting it in the pits of a volcano for a few hours then sending it back to the pit in the future, and melting it down with the minallium alloy to make it stronger and much more flexible. It also allowed me to create an energy field around the device, to protect it from the extreme pressure of space.

The final part is the front casing of the clock, as it was only designed to travel within earths atmosphere. I used the excess glass from the dome, to create a new casing for the atomic clock, carefully melting it around the housing, and soldering it into place. The machine was almost complete, I just needed to update the software, to allow the machine to travel through space, so I connected it to quantum computer, which was linked using an asynchronous algorithm, with the main machine back at the military department. I'm so glad they kept it on, it calculated the size of the update, and the time it would require to initialise it, it then came up with 347 years.

I was like screw that...so I sent it back into the past in cloak mode, inside the sealed chamber at Stonehenge, a one way ticket 350 years in the past. It returned back a few minutes later covered in dust, and smelling of old feet, the update was complete. I sprayed some air freshener coughing back the smell of musty machinery, but there was a catch...the military must have known I accessed their system, so I had to be weary from now onwards, I had to hide it out in Stonehenge, I couldn't live here anymore...

Chapter 32 - Quantum Helix

I see that vision once again of the world in flames, is that really our destiny? I can't let that happen, after all the work I've done to try and prevent this catastrophe from occurring..it feels like with each successive jump through time, I can sense changes in the flow of time around the machine, like it's altering my very DNA.

This is maybe a side effect of all the quantum shifting, as my body has become destabilised from time and space. I attempt to test my newfound strength, now that the changes the alien DNA has made to my body, have stopped almost completely. I first attempt punching one of the concrete walls, that mostly consist of this structure.

My fist goes straight through it, like I just broke a piece of wood...the sensations of pain I'd normally feel are numbed, as the concrete on my knuckles does little to prevent my body healing over the freshly made wounds, what was it doing to my body, was I even human anymore, or something more? I decided to take a blood sample, and looked at it under a microscope, there was traces of the silver substance, from the alien creatures in my bloodstream, along with my normal red blood cells, but there was something else...

I would call them electroloids as there's nothing else I could describe them as, they looked like white blood cells that had become destabilised, or were they just red blood cells, caused by the DNA reversion of travelling through the time stream, over and over again continuously. I could feel a slight energy flowing through my fingertips, sort of like static, but it was barely noticeable, maybe over time it would become more consistent, so I could test the side effects.

I had really bad cramping in my neck, so I took some pain killers, but after three hours they barely had any effect, so I decided to do some exercises with my neck, rocking it from side to side, but that barely worked either, so I decided to take a closer look. I took a mirror, and looked at the definition of my neck just above my shoulder.

The muscles around my neck were bulging out in a way I'd never seen before, was the alien DNA mutating me this soon. I felt suddenly paranoid as to what I'd done, dropping the mirror cutting my foot. I felt a sudden surge of energy, and blasted the chair on the floor in front of me cleanly away, with a shout, like some telekinetic power. The bulging in my neck began to calm down, maybe the side effects of time travel, may help counteract the mutations from the alien genome. I won't know for sure until I've done some more tests.

I took a trip back to Mt Vesuvius before the eruption, with my shirt tied around my waist, my muscles protruded immensely from my body, from climbing this hulk of creation. The embers barely made me break a sweat, as I climbed down onto some nearby rocks, and sat just above the heart of the volcano. I felt like my ears were burning like the very volcano itself, but it wasn't from the heat, it was from the pressure inside my mind...my brain was somehow being supercharged by this newfound energy. I had considered so many things on my journey, but was the most obvious answer the right one, just like Occam's razor...

Chapter 33 - Waking Up

I found myself back at my house once again, I needed to soon make my way to the alien vessel, but one last trip to Stonehenge...I left the railgun here for my own safety. I created a custom holster for the railgun around my waist, after doing some work at the henge, along with one for the sword, so I don't accidentally slice off a finger or two, as it was still incredibly sharp after all this time. I tested the blade against the rocks surrounding this sacred place, honing my skills.

I did this for about twelve days, occasionally resting to eat or sleep, I needed to prepare for what was coming, as they wouldn't just surrender, they would fight to the death...I felt that this time brought me closer to the earth, not just the planet I call home, but the very heart of the planet itself, it's like I could feel nature breathing inside me.

I could hear the tweets, the flaps of a bird wings, the sound of owls in a forest miles away, it's like everything was so much clearer to me now. I needed to head back soon, but first I packed up some emergency rations from my secret stash, and locked away the remains of the ship, from Antarctic wreck.

I felt a strange aura around me, as I walked away from the site, outside of the protection of the nullifying field, I could feel him prying inside my mind... 'What are you doing human, do you really think you stand a chance against us...we are an army, and you? You are but one human vessel, waiting to be crushed under my thumb, like so many others before you. You are walking into a tidal wave, and I am the water that will crush your very soul beneath my weight...

You do not stand a chance...if you give up now, perhaps we might spare you, or we might not...it is your choice after all' I then take a moment to think, then reply calmly and calculated 'I will not stand down, I will not retreat, I will not lie down, and let you beat my race into submission, we are not your slaves, we are free, bound to our will to survive.

You will never take away our humanity, as human beings, we will fight until the very last person, which lives and breathes on this earth is defeated, and so with my dying breath, I honestly hope you are not at the end of my sword, as I shall show you no mercy. As you have shown none, to the people I called my friends, so you too must die like your brethren did, crying and screaming by my blade!'

Chapter 34 - The Price Of Freedom

I make my way onto the ship, but at what cost I wonder...if I do manage to defeat them, will this be the end, I highly doubt it...I sneak my way into the captain's quarters, and there are file cabinets from wall to wall. I rummage through a couple of them, and I find files on almost every person on earth. I wonder how is this possible? How long have been observing us for...

It appears to be a standard office, except for the fact it's on a giant spaceship full of aliens, maybe this is something to do with their immersion with human society, or maybe they're not that much different to us after all...commander...one of the creatures say, as I hide quietly behind the door.

I can hear it shuffle away down the corridor, I look out the door to see where he's going, the cloak is still active on the machine fortunately, so it didn't see it. As I look back over to the right, I attempt to quietly follow him down the corridor, towards the front of the ship...I realise unlike humans, you don't see galleries of their accomplishments, maybe it's because they all look the same, or they're younger than I thought. It could be these are the first of many civilizations that span the galaxy, or it could be these were conceived with the sole purpose of furthering their own existence.

As they were designed differently to human beings, so they would be able to naturally withstand the pressures of space, with their extremely dense gel like structure. It seems to harden when exposed to extreme pressure, as occurred with the corpse I examined...when I submerged it into water. It's body hardened, taking on a crystalline structure, and it's hands formed finned appendages, with a protective membrane forming around it's face.

It seems like even in death their cells can still be activated, much like how real bacteria, can still inhabit the body, long after the illness has expired, such as chicken pox can become shingles...I slowly find my way towards the front of the ship, this place is huge. It's like walking through the carcass of a whale laid on it's back, with the familiar whirring of machinery, attributed in most science fiction shows.

The reality is often much scarier, and dangerous...with strange flashing lights that can be seen through the walls, making sure everything is running correctly, and control panels which open different doors. There are fibre optic sensors, and retinal scanners placed everywhere with giant giant iPads, with lasers attached.

If this were something human's would eventually be capable of, I would be very scared indeed, as we would probably destroy ourselves. I eventually find the front of the ship, and see the same strange creature from earlier, talking to the commander, with a massive viewfinder, showing the upper stratosphere below. I move myself closer to the door, to see if I can hear their conversation, without turning around the larger creature simply states.

'What are your findings?' 'ummm...sir, we have discovered resilience in certain parts of the world, especially the south African region, as well as the Americas, which seem to have been well stocked with firearms, for sometime now. It was before our invasion began, their haven't been any casualties on our side, but...around 3 billion humans have died, as they don't realise their firearms, simply reflect of our flesh, it's natural suicide...'

'hmph he replies simply...this cannot be allowed to continue. If our race is to survive, we must have at least as many survivors, to how many of us there are remaining. If the majority of our race are to survive...our numbers have dwindled since the incident with that...human, who somehow has not been converted by our DNA, it seems some people are immune to it, such as the test subject in Atlantis.

You need to keep me informed about further developments, and no bad news next time...otherwise you will end up like those pathetic creatures who were killed by their own foolish means, be gone!' He scurries away hurriedly as I attempt to hide. He isn't even interested, as I slowly creep towards the larger alien, and draw my weapon towards him. I see him flinch slightly, so he is definitely aware of my presence, so I ready myself for him.

Chapter 35 - The Final Event

'I know why you came here...you seek to destroy me don't you? That is all you pathetic creatures can do...seek and destroy, but it's no matter.' He turns to face me, and continues to say 'We have existed long before your people occupied this planet, it was only a matter of time, before you destroyed yourselves. It wasn't enough to destroy your planet, no no no you wanted to colonise other planets, so you could destroy them as well! You are such foolish creatures' I reply hesitantly 'You don't understand, that wasn't our intention, we needed resources in order to progress, and we are a much more primitive race than you are, unable to harness star's energy.'

We used the natural resources which were available to us, in order to progress to where we are now. You are but a feeble race of fleshy mammalians, we are celluloids born from the bacteria which you would destroy, you would destroy us with your medicines...disease is a natural progress of life, in order to prevent the weak among your kind from surviving, as only the strong can prevail. I shall prove this fact to you now, we are the evolutionary leap you so easily forgotten, now you too shall forget what it's like to survive, your time machine shall not save you!'

I drew the sword from the scabbard, that I retrieved from the pirate ship some 1,200 years ago, which was found amongst the wreckage of a lost vessel, in the captain's own words. I recount that day as if it were from my own memory, as the thoughts flow into my mind, I drift away for a moment or two...

An abomination...how could this have navigated the oceans before us? It is simply a mess of steel, and bubbling hot metal. It burned like the sun, adrift in the ocean, but somehow burning completely immersed in water! The accursed thing had to be destroyed, but first we had to see if there was any loot to gain from it's shell.

We found numerous treasures, but mostly useless things such as scraps of sheet metal, no use for a wooden ship, but at the base of the vessel, I found something of incredible beauty. A blade which shimmered like the stars, I drew it and was almost blinded, it cut through the carcass of the ship like a red hot blade cutting through lead.

It was impossible, this could not exist here...I had to hide it from the crew, only my descendants could learn of this, after all...I did kill the original captain to become captain of this ship, as I knew what lied ahead of me.

That was before I discovered minallium...an unknown substance, otherwise not used on earth. It was as flexible, and strong as titanium, if not stronger, I couldn't test it any lab though, as it was alien in origin. I was as sharp as a diamond etched saw blade, and it suffused light like a gleaming hot rainbow. It was the the only weapon I found which could kill these creatures, as it was created by them, in order to kill those who betrayed them.

The fortunate thing is that this wasn't discovered, otherwise it would have caused irreparable damage to the timeline. This was also one of the largest blades I've found, as they normally took the form of small concealable daggers, but this was a weapon of war, designed to kill. I had the railgun I retrieved from the alien vessel, aimed cleanly at his chest underneath my cloak, his heart wouldn't beat much longer. I had also rigged a kill switch on the time ship, the machine would launch itself directly into the core, loaded with enough minallium to destroy this ship, and the entire neighbouring fleet of vessels with it.

He could sense my hesitation, as he saw my hand across my chest, seemingly stemming the bleeding from a wound I had suffered earlier, but it was not aware of the alien blood which flowed through me, as did it flow through them making much stronger than a normal human being.

Chapter 36 - Endgame

It took me 3 years to find another deposit of this mysterious metal, trapped deep under the ice of the permafrost in Antarctica. I had to travel to the point near the inception of the invasion, before the permafrost had melted enough to extract the alien ship. My memories are hazy after that, as after the explosion at the house, everything went by so fast I can barely remember.

The most intact specimen yet, it must have been a scout ship sent to colonise this continent, carrying a surplus of weaponry, which I reverse engineered to complete my designs for the time ship.

As I found myself back in the present he moved towards me quickly, I shifted slightly to the left, as the quantum shifting took place, I moved a few minutes back in time, my emergency escape procedure. He blathered on about how humans had destroyed this world blah blah blah, but I knew what he was going to do this time, so I ran towards him thrusting the blade forward, causing him to recoil out of the way in surprise, as he reformed from the grey goo.

I said 'game over' he fell for my ploy, as I aimed the rail gun for a clear shot at his head. It whirred to life in a matter of seconds, the minallium round shot forward with such force, it's as if the very air was punctured around it.

As he screamed out in pain NOOOOOOOOOOOOOO! an explosion of goo sent me flying back towards the wall. I quickly got back inside the machine, just as the alarm was sounding, and the neighbouring ships began to fall from orbit, I teleported away sending me rocketing back into the past.

I had destroyed their leader before they could invade, and as I reached the present day, the skies were clear...the forests were green, not reduced to a blackened wasteland like I had seen, I had done it but at what cost? I have still lost everyone I knew, lost forever as an artefact of a forgotten time, so now I must forge a future free of this machine.

I made my way back to that secret place at Stonehenge, and began the process of rebuilding the machine, for a new purpose...as I wasn't going to explore the stars, sitting on my arse for the rest of my life was I!

We were safe...for now at least, the world was safe, and unless if something catastrophic occurred in the next few decades I could live in peace, for a while at least...

THE END

Stockholm 1942

Sir...we are receiving a signal! What is it soldier, hand over the transceiver. He coughs clearing his throat, as he says clearly; This is Stockholm do you respond...I'M COMING FOR YOU ALL! PITIFUL HUMANS! What are you babbling on about, soldier speak up? ugh! He could feel his throat tighten, as the strong force emanating from the receiver, closed his airway and repeated again...I'M COMING FOR YOU ALL! DESTROYER OF WORLDS! I SHALL NOT REST UNTIL YOUR WORLD IS REDUCED TO DUST!...

Quantum Shift before time...

www.ingramcontent.com/pod-product-compliance
Lightning Source LLC
Chambersburg PA
CBHW061441180526
45170CB00004B/1512